BUILDING BLOCKS OF COMPUTER SCIENCE

ALGORITHMS

Written by Echo Elise González

Illustrated by Graham Ross

WORLD
BOOK

a Scott Fetzer company
Chicago

WITHDRAWN

World Book, Inc.
180 North LaSalle Street
Suite 900
Chicago, Illinois 60601
USA

For information about other World Book publications,
visit our website at **www.worldbook.com**
or call **1-800-WORLDBK (967-5325)**.
For information about sales to schools and libraries,
call 1-800-975-3250 (United States),
or 1-800-837-5365 (Canada).

Library of Congress Cataloging-in-Publication Data
for this volume has been applied for.

Building Blocks of Computer Science
ISBN: 978-0-7166-2883-5 (set, hc.)

Algorithms
ISBN: 978-0-7166-2885-9 (hc.)

Also available as:
ISBN: 978-0-7166-2893-4 (e-book)

Printed in India by Thomson Press (India) Limited,
Uttar Pradesh, India
2nd printing July 2024

STAFF
Executive Committee
President: Geoff Broderick
Vice President, Finance: Donald D. Keller
Vice President, Marketing: Jean Lin
Vice President, International Sales:
  Maksim Rutenberg
Vice President, Technology: Jason Dole
Director, Editorial: Tom Evans
Director, Human Resources: Bev Ecker

Editorial
Manager, New Content: Jeff De La Rosa
Writer: Echo Elise González
Proofreader: Nathalie Strassheim

Digital
Director, Digital Product Development:
  Erika Meller
Digital Product Manager: Jon Wills

Graphics and Design
Sr. Visual Communications Designer:
  Melanie Bender
Coordinator, Design Development and
  Production: Brenda B. Tropinski
Sr. Web Designer/Digital Media Developer:
  Matt Carrington

Acknowledgments:
Art by Graham Ross/The Bright Agency
Series reviewed by Peter Jang/Actualize
  Coding Bootcamp

TABLE OF CONTENTS

There is a glossary on page 30. Terms defined in the glossary are in type **that looks like this** on their first appearance.

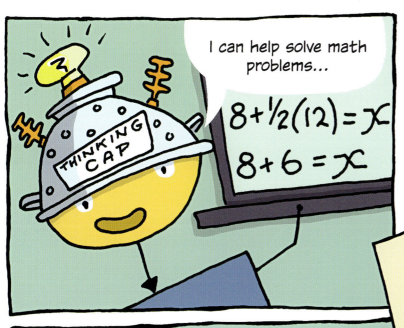

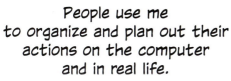

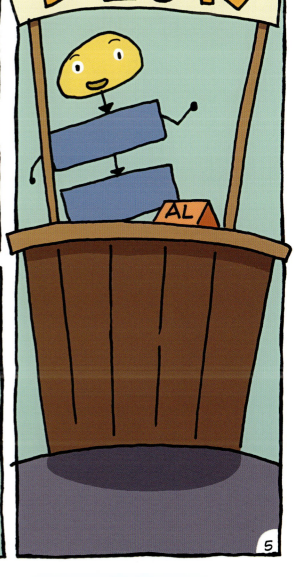

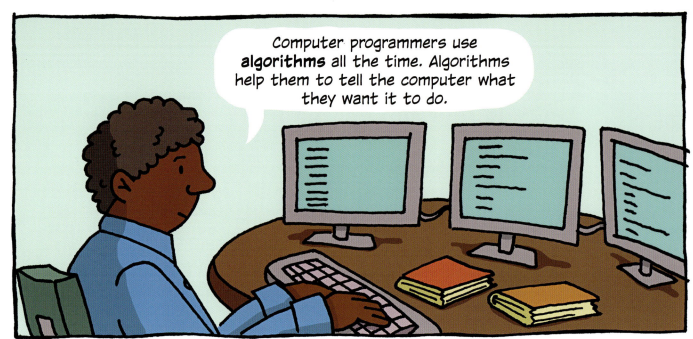

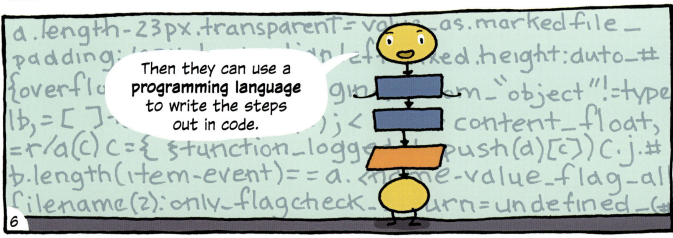

6

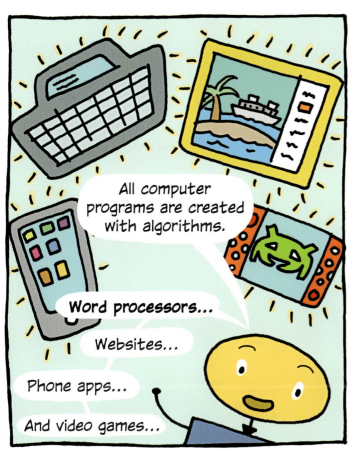

EVERYDAY ALGORITHMS

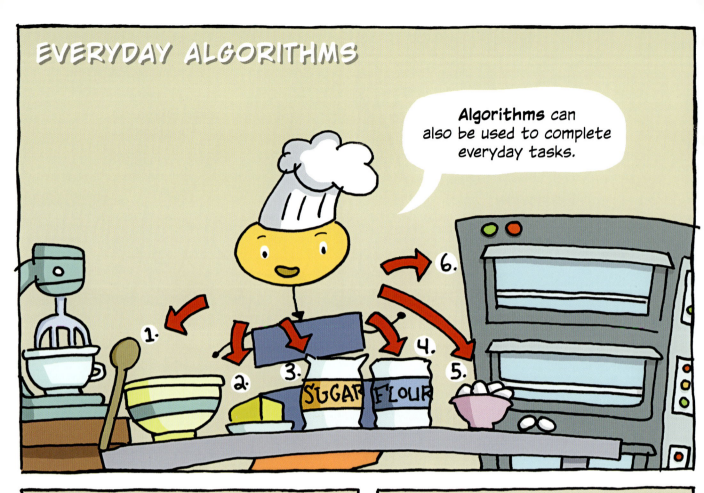

Algorithms can also be used to complete everyday tasks.

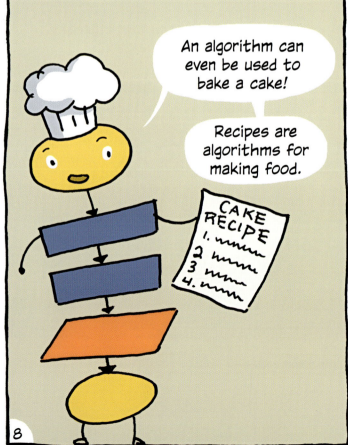

An algorithm can even be used to bake a cake!

Recipes are algorithms for making food.

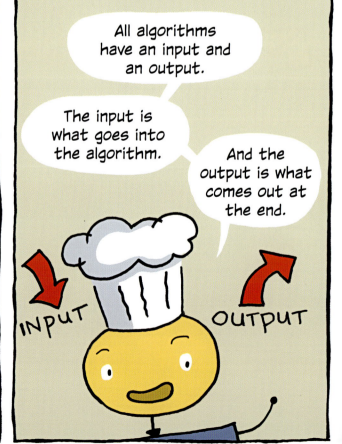

All algorithms have an input and an output.

The input is what goes into the algorithm.

And the output is what comes out at the end.

Programmers write **algorithms** in **code.**

But algorithms can also be written in different ways.

```
temp=cmd_line get_arg
if(temp!="") cont<<

    =cmd_line get arg
if(temp!="") cont<<
cont<<nh,t enter to
cin get( )
```

One way to write an algorithm is by using **pseudocode** (SOO doh code).

In pseudocode, instructions are written out line by line.

Each line describes one step of the algorithm.

Here's an algorithm for playing fetch with a dog, written in pseudocode.

STEP 1: Begin playing fetch.

STEP 2: Throw the ball.

STEP 3: Wait for the dog to return the ball.

STEP 4: Take the ball from the dog.

STEP 5: If you are tired stop playing fetch Otherwise go back to STEP 2.

Another way to write an **algorithm** is to show all the steps in a diagram. This is called a **flow chart.**

Different shapes have different meanings in a flow chart...

An oval shape shows where the flow chart begins and ends.

ANATOMY
of a flow chart

oval

rectangle

rhomboid

arrow

diamond

oval

Programmers also call the oval shape the terminator.

A rectangle represents a process or an action. It is usually the most common shape in a flow chart.

A diamond represents a decision. This shape asks a question and has more than one possible outcome.

A rhomboid (a shape with two slanted sides) represents an input or an output. It shows information that is put into or taken from the computer.

Arrows connect the different shapes. They show the order in which the instructions should flow.

Here's a flow chart showing an algorithm for playing fetch with a dog:

Begin playing fetch.

Throw the ball.

Wait for the dog to return the ball.

Take the ball from the dog.

Tired?

NO

YES

Stop playing fetch.

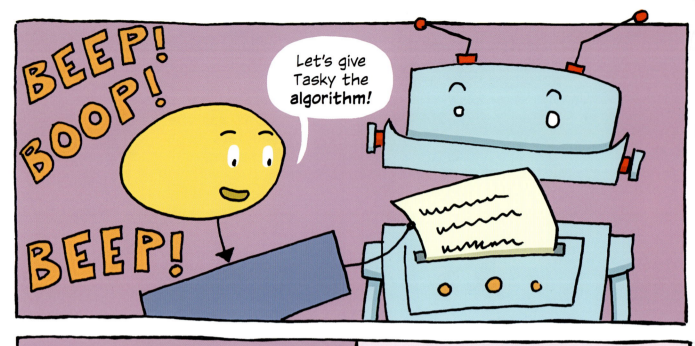

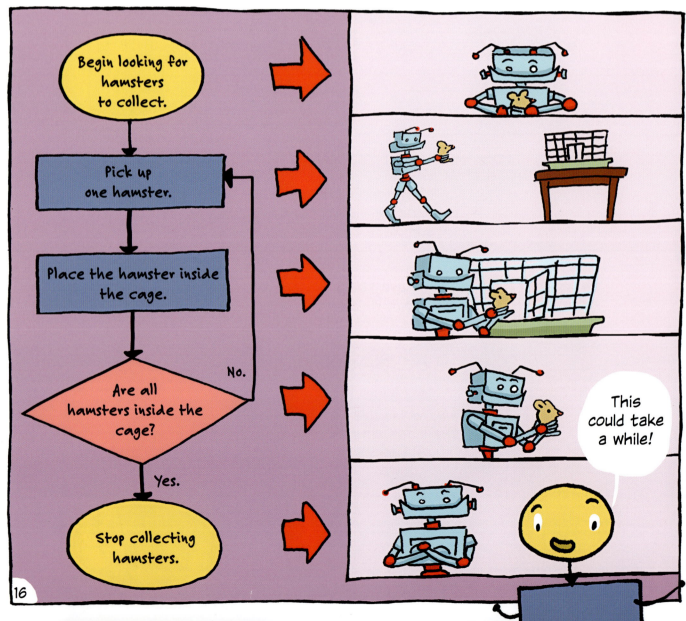

SEARCH ALGORITHMS

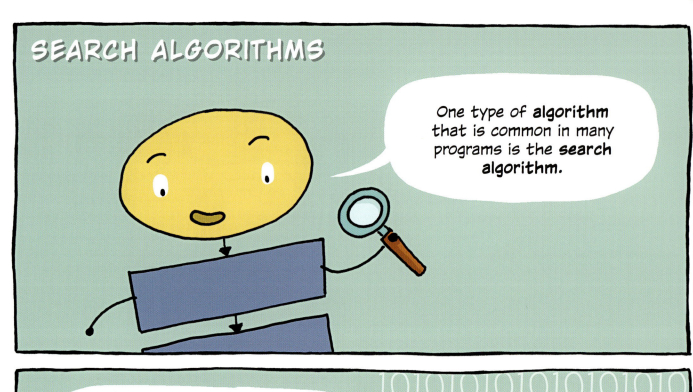

One type of **algorithm** that is common in many programs is the **search algorithm.**

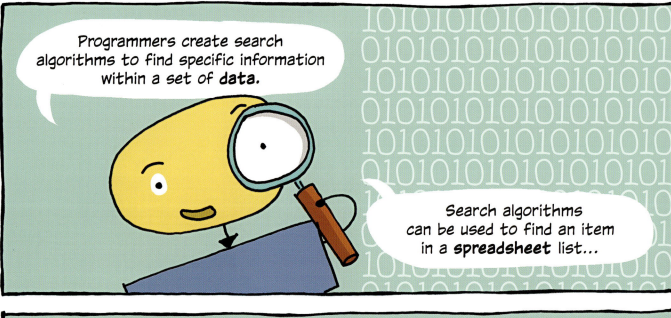

Programmers create search algorithms to find specific information within a set of **data.**

Search algorithms can be used to find an item in a **spreadsheet** list...

To search the internet...

...and to complete many other tasks.

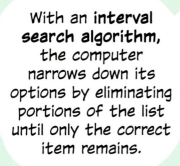

With an **interval search algorithm**, the computer narrows down its options by eliminating portions of the list until only the correct item remains.

Interval search algorithms are also called "binary search algorithms."

They can be helpful for finding an item in a list that's already ordered.

Each of these boxes contains a fruit. The fruits are arranged in alphabetical order. This time, we want Tasky to find the cherries.

The first box Tasky will check is the MIDDLE box.

No cherries in there! But, since "cherries" starts with the letter C, Tasky knows they must be to the left, not to the right.

That means there are only 3 boxes left that could contain the cherries. Again, Tasky will check the middle box.

No cherries yet! But since C comes AFTER B, Tasky knows the cherries cannot be to the left of the banana. So, the cherries must be to the right!

With the interval search algorithm, Tasky did not have to perform as many steps as he did with the **sequential search**.

SORTING ALGORITHMS

SORTING HAT

Another common type of **algorithm** is the **sorting algorithm.**

Computer programmers use sorting algorithms to organize the information in a set of data.

LIST

When you search the internet, a sorting algorithm puts the results of your search in order of relevance, date, or some other factor.

Relevance	Date
1	
2	
3	
4	
5	
6	
7	
8	
9	
10	

AHA!

Sorting makes it easier for you to make sense of the results.

Just about any list of information can be sorted using a sorting algorithm.

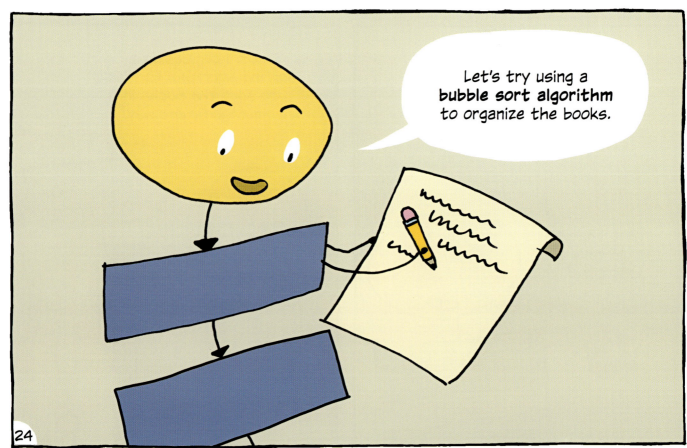

I think I'll try using an **insertion-sort algorithm** to sort the books more quickly.

In an insertion sort, each item is checked one by one, and placed in the correct position.

| 9 | 7 | 6 | 15 | 17 | 5 | 10 | 11 |

| 9 | 7 | 6 | 15 | 17 | 5 | 10 | 11 |

| 7 | 9 | 6 | 15 | 17 | 5 | 10 | 11 |

| 6 | 7 | 9 | 15 | 17 | 5 | 10 | 11 |

| 6 | 7 | 9 | 15 | 17 | 5 | 10 | 11 |

| 6 | 7 | 9 | 15 | 17 | 5 | 10 | 11 |

| 5 | 6 | 7 | 9 | 15 | 17 | 10 | 11 |

| 5 | 6 | 7 | 9 | 10 | 15 | 17 | 11 |

| 5 | 6 | 7 | 9 | 10 | 11 | 15 | 17 |

ALGORITHMS IN HISTORY

Algorithms are an important element of computer programming...

But they've actually been around since ancient times!

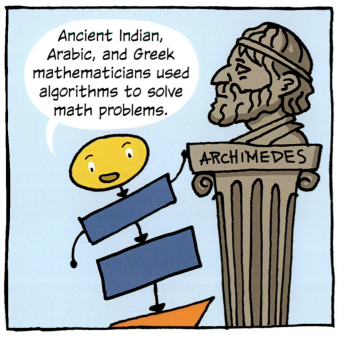

Ancient Indian, Arabic, and Greek mathematicians used algorithms to solve math problems.

ARCHIMEDES

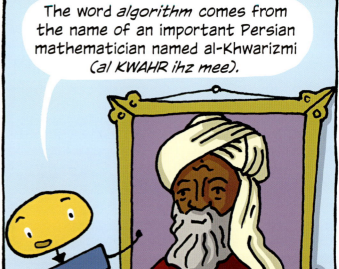

The word *algorithm* comes from the name of an important Persian mathematician named al-Khwarizmi *(al KWAHR ihz mee)*.

Al-Khwarizmi helped spread modern mathematical concepts from ancient Indian and Arabic writings to the rest of the world.

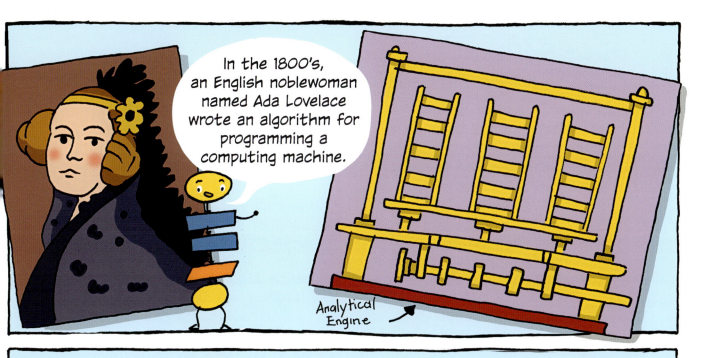

GLOSSARY

algorithm a set of step-by-step instructions used to write computer programs. Algorithms are also used to solve math problems and other problems.

bubble-sort algorithm a sorting algorithm that tells the computer to put a list of data in the correct order by arranging the data in pairs.

code instructions written in a programming language.

data information that a computer processes or stores.

flow chart a diagram that uses shapes and arrows to show the order of steps that make up a process.

insertion-sort algorithm a sorting algorithm that tells the computer to check each item in a list of data one at a time to put the list in a particular order.

interval search a search algorithm that tells the computer to eliminate chunks of data from a list to find a particular piece of data.

programming language a set of symbols and rules that programmers use to write computer programs.

pseudocode a description of a computer program written in human language.

search algorithm an algorithm that is used to find specific information in a list of data.

sequential search a search algorithm that tells the computer to check each item in a list of data one at a time to find a particular piece of data.

sorting algorithm an algorithm that is used to put a list of data in a particular order.

spreadsheet a document in which data is arranged in a grid.

GO ONLINE

Ready to try making and following algorithms yourself? Go to this website and click on the Algorithm Origami activity to make an origami penguin! You'll find all kinds of fun computer science activities! Click on the Robot Friend activity to find an algorithm game you can play with a pal.

www.worldbook.com/BuildingBlocks

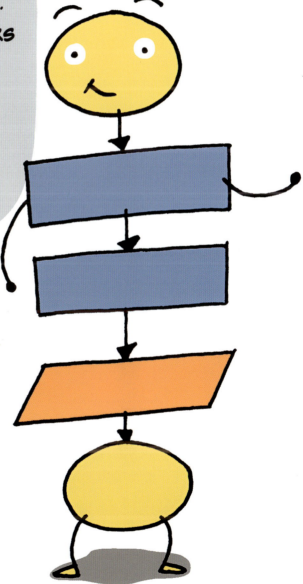

INDEX

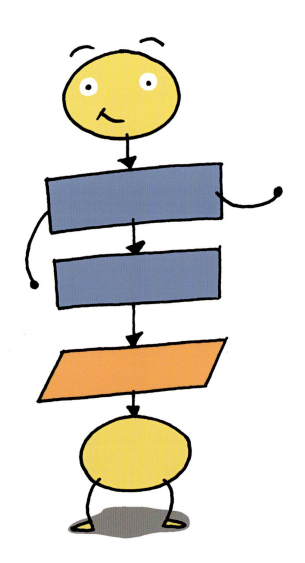